What the Moon Is Like

Here is concise, reliable information that provides a firm background for understanding our exploration of the moon.

How large is the moon? Is there a man in the moon? What are the dark parts made of? The light parts? How much would you weigh on the moon? How would you move? These, and many questions like them, had to be answered before people set foot on the lunar surface.

What the Moon Is Like gives the young reader some of the answers. Even more, with simple words and vigorous pictures, it gives a sense of the desolation and barrenness of Earth's natural satellite. Most of all, it gives the reader an understanding of our need to explore, our need to find out about this part of the universe around us.

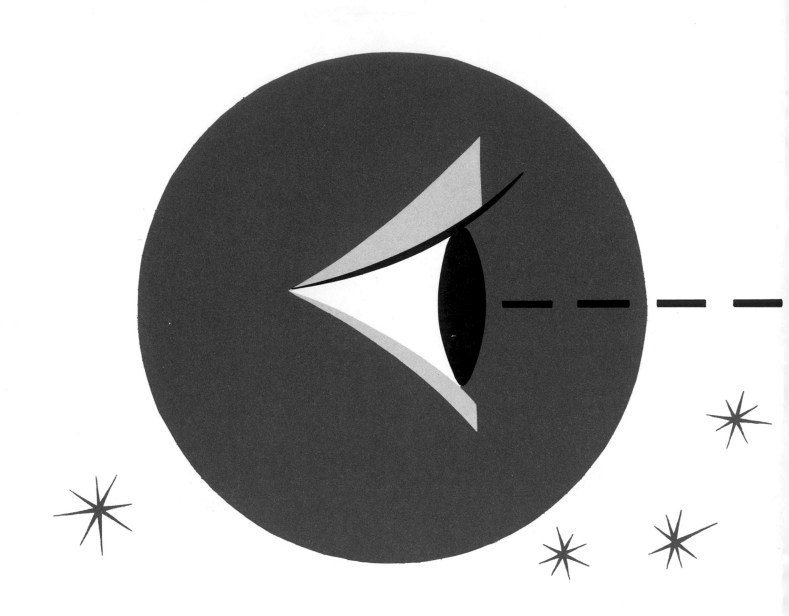

What the Moon Is Like

by Franklyn M. Branley

Illustrated by Bobri

A Harper Trophy Book

THOMAS Y. CROWELL   NEW YORK

LET'S-READ-AND-FIND-OUT BOOKS

Let's-Read-and-Find-Out Books are edited by Dr. Roma Gans, Professor Emeritus of Childhood Education, Teachers College, Columbia University, and Dr. Franklyn M. Branley, Astronomer Emeritus and former Chairman of the American Museum–Hayden Planetarium. Text and illustrations for each of the more than 100 books in the series are checked for accuracy by an expert in the relevant field. Other titles available in paperback are listed below. Look for them at your local bookstore or library.

A Baby Starts to Grow

Bees and Beelines

Birds at Night

Corn Is Maize

Digging Up Dinosaurs

A Drop of Blood

Ducks Don't Get Wet

Fireflies in the Night

Follow Your Nose

Fossils Tell of Long Ago

Hear Your Heart

High Sounds, Low Sounds

How a Seed Grows

How Many Teeth?

How You Talk

It's Nesting Time

Ladybug, Ladybug, Fly Away Home

Look at Your Eyes

Me and My Family Tree

My Five Senses

My Visit to the Dinosaurs

No Measles, No Mumps for Me

Oxygen Keeps You Alive

The Planets in Our Solar System

The Skeleton Inside You

The Sky Is Full of Stars

Spider Silk

Straight Hair, Curly Hair

A Tree Is a Plant

Water for Dinosaurs and You

Wild and Woolly Mammoths

What Happens to a Hamburger

What I Like About Toads

What Makes Day and Night

What the Moon Is Like

Why Frogs Are Wet

Your Skin and Mine

What
the Moon
Is Like

Did you see the moon last night?
Was it big and round?

When the moon is round, people say they can see "the
man in the moon."
The dark and light parts make them think of a
mouth, a nose, and two eyes.
That is why they say there is a man in the moon.

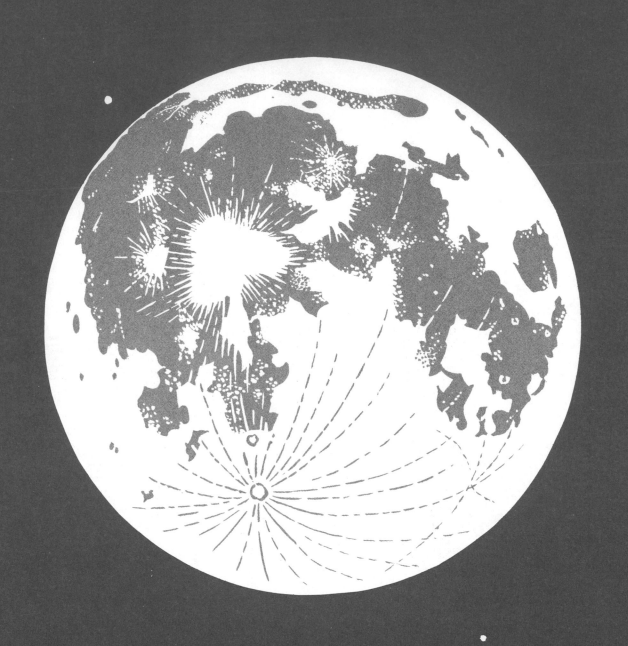

Next time the moon is big and round, look at it.

The moon is far away, but we can see it very well.

We see that the moon is round like a ball.

Parts of the moon are bright. Other parts are dark.

Craters look bright. Seas look dark.

These are the bright and dark places you see when
you look at the moon.

Make believe you are on the moon. What would it be
   like?
There is no air on the moon.
You cannot live without air, so you would need a
   space suit on the moon.
There would be air inside the space suit.

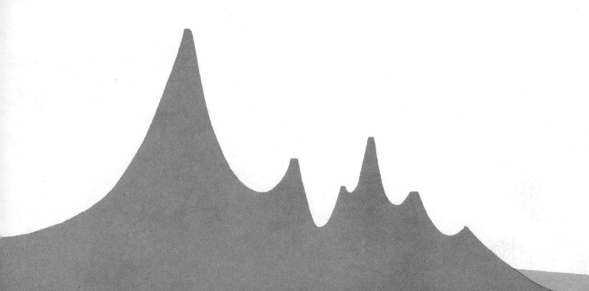

If you were in sunlight on the moon, it would be very hot. The space suit would keep you cool.

When you were out of the sunlight, it would be very
   cold.
The space suit would keep you warm.

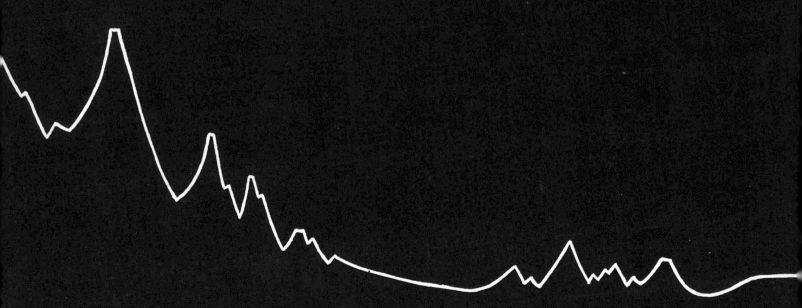

Do you weigh 60 pounds?
You would weigh only 10 pounds on the moon. So
would your friend. You could pick him up easily.

You could pick up big rocks. You could throw them far, too.

You could jump from place to place.
You could move easily in your space suit. You could
jump right over a house, if there were houses on
the moon.

You could explore the moon mountains. They are
   high and steep.
You could look into deep valleys. You would see
   rocks all over.
You could look into deep cracks in parts of the moon.
In the cracks you would see only rock and stone.

Nothing lives on the moon.

You would see no plants, flowers, birds; no grass, no animals.

You could explore the moon craters.
Craters are flat places with hills around them like a
   wall.

Some craters are little. You could walk across them.
Other craters are big. One crater is one hundred
   eighty miles wide.
You could jump up the wall of a crater to see the
   other side. You would see more rock and stone.
   Maybe you would see a deep layer of dust, so deep
   that you would sink into it. Maybe you would see
   more craters.
Maybe you would see a lunar sea.

Lunar seas are big flat places on the moon.
We call them seas because they are flat, not because
they hold water. There is no water on the moon.

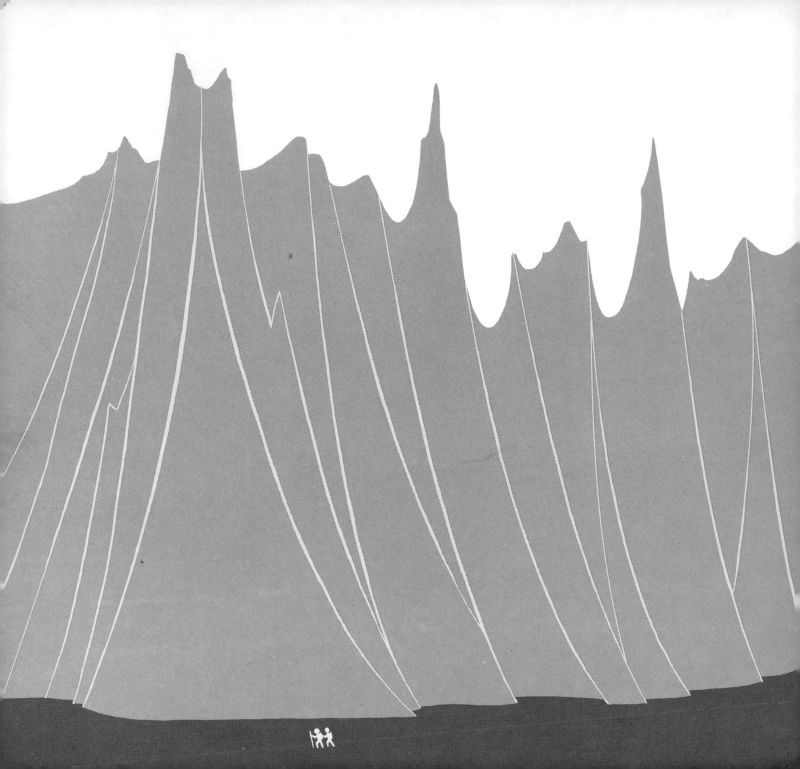

Some of the seas have high walls.
If you stood beside them, you would look very small.
Do you see the boy and girl in the picture?
That is how small you would look.

We want to know more about the moon.
We send rockets there to gather information and
    send it to us.

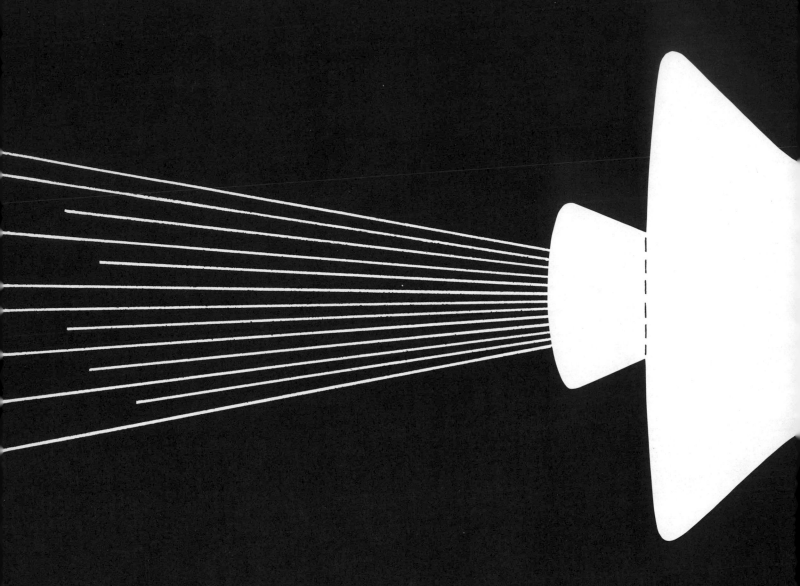

We send men to the moon.
They live inside the rocket that takes them to the moon.

Outside the rocket the men wear space suits. The
space suits keep them from getting too cold or too
hot.
They take tanks of air with them so they can breathe.
They take water and food, too.

When men go to the moon, they explore the mountains, the deep cracks, the craters, and the seas.

They are moon explorers.

Moon exploring is exciting.
It is an adventure.
Some day you may be a moon explorer.

About the Author

Franklyn M. Branley, Astronomer Emeritus and former Chairman of the American Museum-Hayden Planetarium, is the author of more than 50 books about astronomy and other sciences for young people. He is also coeditor of the Let's-Read-and-Find-Out science books.

Dr. Branley holds degrees from New York University, Columbia University, and the State University of New York College at New Paltz. He and his wife live in Woodcliff Lake, New Jersey.

About the Illustrator

Vladimir Bobri is equally at home—and well-known—in musical and artistic circles. He is president of the Society of the Classical Guitar, editor of *Guitar Review* magazine, and an acknowledged authority on gypsy music.

Mr. Bobri was born in the Ukraine where he attended the Kharkov Imperial Art School. He earned the money to come to this country in 1921 by making decorations and costumes for the Ballet Russe of Constantinople. Mr. Bobri has received a number of awards for children's book illustration as well as many citations from the Art Directors Club for his advertising design.